Living UFO's

The Ocean Above Our Heads

Art Credit: Other Side Prints

A Study of Atmospheric Biology

Written By: Justin England

Living UFO's
The Ocean Above Our Heads
A Study of Atmospheric Biology

Printed in the United States of America

First Printing, 2024

ISBN 979-8-218-53347-2

Cryptids of the Corn Podcast
P.O. Box 75
Ada, Ohio, 45810
www.cryptidsofthecorn.com

Ordering Information:
Quantity sales. Special discounts are available on quantity purchases by corporations, associations, and others. For details, contact the publisher at the address above.

Table of Contents

Foreword:

Written by Alexis Fiden, a graduate level biologist with a focus on animal behavior and cryptozoology.

As a biologist, I always look for the reasons “why” a cryptic creature may exist and how it may be possible without humans being aware. Justin has done a phenomenal job meshing cryptozoology with the traditional study of biology to explain just how living UFO creatures could exist in our upper atmosphere. His speculative biology gives excellent insight into each encounter individually, taking time to explain why each sighting has been significant. Justin has taken time to do his due diligence with research both in the UFO encounters as well as oceanic biology to find mirroring trends. After reading the book, I am convinced that Justin is onto something, and I hope this sparks additional research into life in the upper atmosphere. I cannot wait to read more!

Dedication

To you, Atlas, may this remind you this world is a wonderfully strange place with mysteries around every corner for you to explore. Your dad loves you and can't wait to see what you discover.

"My job, my mission, the reason I've been put onto this planet, is to save wildlife. And I thank you for comin' with me. Yeah, let's get 'em!"

-Steve Irwin

Thank You!

Thank you to my amazing wife Emily, for always being there to support me.

Thank you to my Co-host Jay for being on this crazy ride with me.

Thank you, famous author, Michael Thompson, for taking the time to walk us through the process and producing amazing art for the book.

And thank you to our amazing editor Alexis Fiden, for taking your time and skills to work with me on this amazing project.

About the Author

Justin England, aka Mr. E from Cryptids of the Corn Podcast was on a fisheries crew (field biologist) that did endangered species work and water quality assessment work throughout the Midwest. He is a full-time husband and father of 3, works as a 4H advisor and is head of the county poultry department. With this experience comes a lifetime of love and passion for nature and animals, heavily influenced by the great Steve Irwin. Justin has an encyclopedic knowledge of the Animal Kingdom and uses this gift to make educated speculations in the field of cryptozoology. All this coming from a grounded point of view that's open to all opinions, allowing you to make up your own mind about what is truly possible here on Earth. This unique blend is a perfect storm of education and entertainment that you can find in anything Justin has put his efforts into. - Jay Wolber, Cryptids of the Corn Podcast Co-Host

Goal

The goal of this book is to not have a complete understanding of this amazing phenomenon because there's so much, we still do not know. Instead, this is just an introduction to some of our theories of our gigantic organic neighbors and using speculative biology we can make educated guesses upon the nature of these fascinating creatures. Hopefully this book will get you all thinking in gaining your own theories into the world above.

Introduction

Our skies have been abuzz with mysterious flying objects for millennia. Some of the strange objects have been seen shooting fire, abducting cows, or being piloted by terrifying entities, but some of these UFO's have a more organic feel to their behaviors. These organic UFO sightings represent a small fraction of total UFO encounters every year but are no less important. These creatures seem to have a far more home bounded origin than your traditional space aliens. The sky above our head may have more in common with our open oceans' environments than any other place on the planet. NASA studies revealed thousands of species of microbes and planktonic life in the stratosphere that mimic the ocean. It would seem we are only missing the whales and the sharks… or are we?

The Upper Atmosphere and its Similarity to Open Ocean Environments

The open ocean and the upper atmosphere have much more in common than one may decipher upon first look. Both environments have giant open spaces, pressure extremes, temperature variances, a lack of hard structures for life to form on, and both may suffer at the mercy of giant storm systems. When we see other environments in the world that have this much in common, we see a phenomenon of convergent evolution. Convergent evolution is when two non-closely related species evolve to look very similar. This occurs when two different environments utilize similar body plans to fulfill similar environmental niches. Convergent evolution will be a common theme when we look at the speculative biology of the upper atmosphere and the things that have been observed living there.

Layers of the Atmosphere

The layers of the atmosphere are huge and unique. The first layer is the troposphere, this is the layer we are living in. The troposphere ends about ten miles above the ground depending on your local altitude. This layer has all the Earth's commercial air traffic and almost all private and government air traffic as well.

The next layer up is the stratosphere, this begins about ten miles up and extends its reach up to thirty miles above the ground. This layer is home to the ozone. The stratosphere's weather, for the most part, is stable with the temperature variances being very consistent with the rise in attitude. This means there are very little weather effects in this layer. The only problem is, if a flying creature were to get too close to where the stratosphere and the troposphere meet, a storm system could suck the creature down into our layer.

The next layer is the mesosphere. This layer starts about thirty miles above the surface and ends somewhere around fifty miles above the ground. This layer is very important to Earth safety due to it burning up most of the meteorites that enter Earth's atmosphere. This layer is also the exact opposite for temperature variances, as it gets cooler the higher in altitude you go.

The next two layers are the thermosphere and the exosphere. These layers are of very little interest in our study of life in the upper atmosphere, due to the gasses in these layers being separated out by density instead of being mixed together. It would theoretically not be impossible for life to form here, but it would be extremely unlikely, and even more, implausible to form large life. The altitude of interest to us is from twenty-five miles to forty- five miles above average ground level. This is the area in the atmosphere around where the stratosphere and the mesosphere meet. This area has the highest temperature above the troposphere while still having a mixed atmosphere. This area has very stable weather and temperature conditions, and still has ice clouds to gather water from. This specific area would be a perfect area for complex life and integrated food webs to develop.

Layers of the Atmosphere

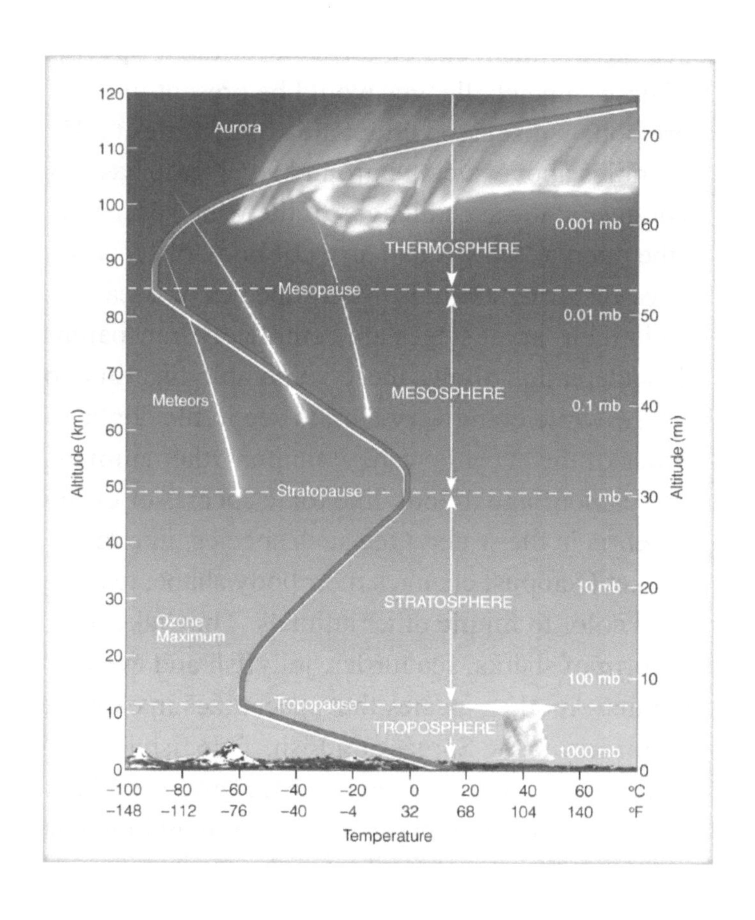

Open Environments

The open ocean presents species with many challenges. To adapt to an environment with nowhere to hide, take shelter from storms, to get out of the sun's harsh rays, and to find food in a vast open environment can be difficult for many species. These same challenges would be present in our previously discussed target area of 25 miles to 45 miles above average ground level. Life has conquered these challenges of the wide spaces of the open ocean with an array of body plans and awe-inspiring abilities. Many species of oceanic fish in their larval stages are completely transparent. This adaptation allows these vulnerable fish time to grow while being very hard to see in the “light” zones of the ocean’s water column. Other animals use camouflage to look like toxic species or even debris in the water. One such species, like the Blanket Octopus, changes their body shape, texture, and color to mimic other animals. They take the form of sharks, sea turtles, jellyfish and even humans. Another species that specializes in extreme mimicry is the Sargassum Fish. This fish is so specialized in mimicry, many aquarium goers think their tanks are completely empty as it may look just like the plants in their habitat.

Another very popular survival adaptation is extreme speed. These animals are often extremely streamlined and able to accelerate to high speed very quickly. Flying fish use speed, as well as their gliding ability, to evade predators. These fish can reach speeds of 43 mph and then jump out of the water and glide one there wing-like fins for great distances. Life forms in the upper atmosphere have probably evolved similar abilities to conquer similar problems in an open environment.

We have reports of organic UFO's being almost completely transparent. Even very large Manta Ray like creatures are observed as almost completely transparent. Eyewitnesses at close range have seen the stars at night through these creatures. We also have reports of them using camouflage to disguise themselves as clouds. These clouds are very convincing until they move against the wind, dodge obstacles and are seen eating and producing biological materials. Living UFOs are also capable of incredible speed. Our silver dice-like UFOs are observed going slow followed by instant acceleration to sonic speeds. They're also witnessed to "play" with aircraft like a dolphin playing with a slow tugboat.

The similar environment of open oceans to the open sky have created similar adaptations and body plans to fill in these incredible ecosystems.

The Extreme Pressure

Pressure is very similar to temperature in the mesosphere. There is a steady decrease in both temperature and pressure with the rise in altitude. The measuring units of atmospheric pressure are called millibars. The troposphere, where we live, ranges from two hundred millibars at higher altitudes to one thousand millibars at sea level. The target area between the stratosphere and the mesosphere has a pressure of 1 to .01 millibars. This low pressure would allow living creatures with gas filled chambers to be able to maintain lift. However, if they were closer to surface level they would potentially struggle or even die. As you'll see later in the book, many of these beings struggle and seem to be in very poor health when they are too low in altitude, with some even having witnessed to die. We see this exact phenomenon in some deep-sea animals as well. The famous Blobfish in its natural habitat looks nothing like that of the mess of flesh it becomes when it's pulled to the surface. This is caused by depressurization. The fish's body is designed to deal with the pressure extremes, so, when brought up, the lack of pressure causes their body to deform. However, this is not the case for all deep-sea life. Life has evolved all kinds of ways to get around pressure extremes and transiting to pressure extremes. This is especially true when going from low pressers to much higher pressers. Sperm Whales have evolved to be expert pressure survivors. They dive straight down, to depths that would crush a normal submarine. They are able to survive such extreme fluctuations in pressure by having flexible bodies and bones, lungs that can collapse so they don't explode, and the ability to store oxygen in the body's tissues instead of their lungs. All of these adaptations work together to allow them to hunt Giant squids in 2000-meter-deep water. Creatures in the upper atmosphere may have evolved similar adaptations to help "dive" to the troposphere in search of food or maybe even to seek the possibility of reproduction. They would likely have very flexible bodies, able to constrict with presser extremes, allowing them to become more dense to help maneuver in this different environment. Another evolutionary adaptation they could have would be an organ similar to that of a fish's swim bladder. The swim bladder allows fish to maintain buoyancy and rise and lower in the water column by producing gasses and releasing them like a hot air balloon. Nature has already conquered the complex issue of pressure extremes and pressure changes. However, these entities may have completely different mechanisms to get over this biological hurdle that we can't even theorize.

Temperatures

The layers of interest (Stratosphere and Mesosphere) have a wide range in temperature depending on altitude, but on average, temperatures range from as low as -40°F to as high 35°F. The stratosphere has an incremental increase in temperature with the rise in altitude until it reaches the meeting point of the mesosphere. The mesosphere has the opposite effect, where it gets incrementally colder as the rise in altitude increases. The meeting point of these two layers of the atmosphere has the highest temperature, allowing for gasses and liquid water to form, giving us the best opportunity for life to form within this complex ecosystem. Many lifeforms, large and small, call this temperature range home. From the mighty Bowhead Whale to the largest schools of krill on the planet, many animals thrive within this same range of temperatures in the ocean. These temperature ranges have a large misconception that they are incapable of supporting complex food webs, but research in the Arctic Ocean and under the permanent ice sheets share with us a very different story. Due to seasonal food abundance, these areas produce biomass in the billions of pounds and are the birthplace of the many species of whales. Biomass is the combined weight of all life in a system added together. The upper atmosphere is most likely no different than the Arctic Ocean, with the largest percentage of biomass being processed as plankton and algae. These small creatures and plants are found in the upper atmosphere, which we will cover in a later chapter. These cold temperatures could be another factor in why these creatures may be "diving" into the troposphere as they may need to warm up on occasion. Humpback Whales do this when traveling in cold currents. They will swim into shallow, warm waters, to relax for a bit. They could not survive in this environment for long, but it's a nice place for them to rest every once in a while. These temperatures were once believed to completely limit the complexity of life that would be able to exist in the higher layers of the atmosphere, but, with advancing technology, we are discovering new species in the atmosphere every day.

Lack of Hard Structures

A lack of any hard or fixed structure is a biological factor that many open ocean life forms and upper atmospheric creatures have had to deal with. In the open ocean, this presents problems avoiding predation, as well as with reproduction and offspring care. Like we discussed in previous sections, animals in the ocean have evolved many ways to avoid predators with nowhere to hide. These animals use camouflage, transparency, and speed to avoid predators. Another thing some of these ocean animals may do when predators are after them is to dive into the ocean's depths at high speeds, or head for the nearest oceanic structure, like reefs, rocks, or the bottom of the ocean depending on depth. Reproduction is easily the most difficult obstacle to overcome. For some fish, spawning can be quite difficult due to mate selection and location, or the ability to find each other. To overcome this obstacle, many of these animals have specific spawning or breeding locations. Some fish, like tuna, will travel to the same area over a rocky outcrop or an ocean mound to release eggs and sperm as a large group, which is called scatter spawning. Some other fish, like salmon, travel upstream to freshwater nurseries to fight for mates and leave eggs and larval fish in a safer environment before the offspring are old enough to return to the ocean. Some ocean dwellers, like Great White Sharks, use sexual reproduction. This means a male and female need to meet to produce offspring. They most commonly breed around large group feeding sites, like a whale carcass. Males and females select a mate and breed, then after some time the female will give birth to live pups. Sea turtles have also come up with ways to overcome the challenges of the open ocean by coming on to dry land to lay their eggs in the sand. The turtle has a very difficult time on land, and many die in the process but their eggs need air for the incubation process. This makes the risk necessary for the mother to secure the next generation. Creatures in our upper atmosphere would face similar challenges and would likely develop similar solutions because their environment and niches would be almost identical. Something similar to scatter spawning could be effective if their "eggs" could drift on air currents until they hatch. This system would have less mate selection and likely never affect the ground miles below. If the "eggs" are caught in a down draft of wind that could make it all the way to the ground, this could explain phenomena of things like black snow and other similar events. Sexual reproduction would be the easiest option for a creature that produces less offspring or that could maintain altitude right after birth. This type of birth is used by whales, dolphins, and some sharks, amongst other creatures. In the case of whales and dolphins, they also use advanced parenting techniques, so their offspring don't immediately need to fend for themselves. Sharks, on the other hand, have no parenting so their offspring can hunt and defend themselves right after birth. Sexual reproduction is used mostly by apex and predatory species in the ocean. This type of reproduction may be used by some atmospheric creatures that fit those same roles. Traveling to a spawning area or nursery area would be another technique commonly used by these entities. This could explain why some of these creatures are seen low to the ground and even seemingly writhing in pain. These life forms could be risking themselves to secure the future of the species by going to the ground to lay eggs or release offspring. This would explain why the Manta Ray UFO's are often seen very low to the ground, near remote areas with food and water sources nearby. This could also explain some smaller Manta Ray-like creatures we will discuss in later chapters. The lack of structures available for shelter and reproduction is a hard environmental factor to overcome, but, done again and again by nature, this could be why we are having run-ins with these amazing creatures.

Life is Proven in the Upper Atmosphere

NASA and many other organizations have, in recent years, begun a survey of microscopic life in the upper troposphere and lower stratosphere. Up until the last ten years, many scientists believed the upper atmosphere was barren for most forms of life due to the extreme nature of the habitat, but these surveys suggest the exact opposite. The earth is full of a group of animals called extremophiles; these animals fill in niches we would conventionally think were impossible to fill. These animals range from crabs and tubeworms that feed off of boiling sulfuric vents to tardigrades that can survive in the vacuum of space. The conditions in the upper atmosphere have already been conquered by a diverse group of life that scientists have identified in these studies.

The studies and works we will mostly be focusing on will be that of David J. Smith and Samantha M. Waters of the Space Biosciences Research Branch, NASA Ames Research Center. The goals of these projects were to study the earth biomass that comes up to the troposphere and enters the stratosphere and how this lower atmospheric life handles the radiation and dryness of the space. The study was used to see what life may survive on Mars due to environmental similarities. They found microbes, fungi, and bacteria in droves in the atmosphere. This study shows that not only can surface bound life handle these extremes, but some life has adapted to completely live in the upper atmospheric conditions. Other studies have looked into the number of species found in the stratosphere and have found numbers in the thousands representing most clades of microbial life.

These studies also show that a large amount of troposphere (surface level) biomass transfers up to the stratosphere naturally and regularly. This biomass is mostly organic dust but also includes larger lifeforms like fungus and bacteria. The largest amount of surface biomass being pushed upward occurs around mountain ranges and storm systems. Mountain ranges act as a natural weather barrier and cause large wells of water and biomass to be pushed into the stratosphere. Storm systems also push biomass and water. They do this through the pressure zones and vortexes they create, acting like a waterspout, throwing things into the atmospheric levels above. This natural phenomenon also mimics the ocean and how currents push nutrient rich water to certain locations to create large feeding areas that produce massive amounts of aquatic biomass. Organic UFO's are often seen in concentration around mountain ranges and right before or after heavy storm systems. This could be due to them filtering out this biomass as a food source, similar to whales hanging out in the areas with the heaviest food concentration.

Planktonic-like Life and the Speculative Biology of the Food Chain Above

As discussed in previous sections, the upper atmosphere and the ocean continue to have revealed similarities. While the upper atmosphere is lacking true planktonic life, it has a close aggregate in the many species of fungi, algae, bacteria, and their endospores, both from surface level upwell and from species found only in the atmosphere. While most of these kinds of life are smaller than true planktonic life, they can serve the same function, which is to feed smaller species of creatures and huge filter feeders to yet be classified. Species will often fill available niches if a food source is not being taken advantage of. The available food production for this system is hard to measure due to the multitude of input sources. The uptake of surface level biomass comes from all our oceans, lakes, rivers, rainforest, plains, and the yet to be determined production level of the natural atmospheric life itself. These levels of food production reveal a large gap in this newly discovered food chain. Nature does not often leave any blank space empty, even in the most extreme environment. Complex food webs will develop from deep sea vents to hot spring pools. It only makes sense that with the abundance of bottom-trophic level food, there may be a larger creature to feed on them. This is where organic UFO's come in. These living creatures have been seen as forms that fit with open ocean counter parts, if not just on a larger scale, everything for squids, eels, fish manta rays and some even more foreign body designs. These beings are often observed to be massive, sometimes being hundreds of feet wide and a mile or so long. In open environments such as the ocean, large body plans often are favored such as the Blue Whale. Due to the similarities of these environments, it stands the similar principles of biology follow, such as open ocean gigantism, there would be upper atmosphere gigantism. There are several reasons open spaces breed bigger creatures. Firstly, Kleiber's Law.

Kleiber's Law is a biological principle that, in short, states that the larger an animal is, the more energy efficient it is. Kleiber's Law works due to food supply. If the food supply is scarce or there are long distances in between food sources, a large body can hold more food reserves until a meal is found again. A large body also takes less energy to move greater distances than that of a smaller creature. Another principle that may go into effect in environments like the upper atmosphere is Bergmann's Law. Bergmann's Law states that a decrease in temperature can cause deep-sea gigantism. These creatures grow much slower, grow larger cells, and live much longer. The increase in body size helps conserve heat in the depths where temperatures can get below freezing. Since conditions in the upper atmosphere have a similar temperature range to that of the ocean floor, this could be another factor leading to these sky creatures being so massive.

Speculative Biology and Encounters of Different Organisms of the Upper Atmosphere

There have been many encounters of UFO's that have been believed to be organic in nature. These creatures may be earthbound life that have evolved to take advantage of the environment the upper atmosphere has to offer. These next sections will show speculative Biology of each category of atmospheric beasts, and encounters with each different organism type of the upper atmosphere. This will include the date, location, witnesses if any and the encounter details. This will also include a speculative biology analysis of each sighting which will include biology, behaviors, and relevance to other encounters.

Convergent Evolution

These categories are named for their earthbound look-alike. This is in relation to a phenomenon in nature called convergent evolution. Convergent evolution is where similar environmental pressures shape different types of life into similar body plans. This occurs when a similar body plan would be the most efficient for both creatures in their niches.

For example, sharks and dolphin being vastly unrelated but have extremely similar body plans.

Where Are They on Radar?

These creatures are often seen from the ground, around planes and jets. Some have even been seen by the pilots themselves, who show us close ups of instruments in the vehicle, but nothing shows up on ground radar. This same radar phenomena can be seen with sonar in the ocean. Large organic beings, like whales but do show up close, even despite their massive size. This is a result of low density. Long distance sonar and radar are designed to pick-up high-density objects like planes, jets, and missiles from long distances, but low-density organics would be almost invisible. Close range sonar and radar has a much easier time picking up organics, which is how whaling vessels used to track down whales to hunt. The phenomena of pilots being able to pick these creatures up on their instruments, but the ground tower cannot. That is exactly what we would expect from living creatures around jets and planes.

Manta Ray

Speculative Biology:

A group of sightings collectively looking like a well-known sea creature may fill the niche in the upper atmosphere that whales fill in the ocean. Their appearance is superficially that of a Manta Ray. Differing from their oceanic counterparts, this group is often humongous, being reported to be bigger than a 747-jetliner, with close to translucent skin. They have been seen slowly flapping their "wings", but this motion has little effect on the area around them. The flight and ability to "float" is likely due to the use of an organ similar to a fish's swim bladder. These creatures are using "wings" to steer while using their bladder to rise and lower in the air. This is evident in their slow and peaceful motions. Their behavior reminds us heavily of an open ocean filter feeder. They have been known to have large bioluminescent organs on their bodies, giving them the appearance of having "lights" on their body. These "lights" can lead to them being confused for materialistic aircraft until the observer gets the chance for a closer inspection. They are often doing large figure eight motions, which is a very efficient way to feed for the slow animal. We believe these animals are using the same techniques as their oceanic counterparts to filter food out of the air column. Encounters with these behemoths are often related to storm systems and air pressure fluctuations. These creatures seem not to pose a direct threat to humans and may not notice our presence at all.

Art Credit: Michael Thompson

Sightings

Title: Manta Ray Over the Ohio River

Location: Mason County, WV

Date: Dec. 3rd, 2003

Witnesses: Unnamed Couple and Unnamed Mother and Daughter

Encounter: In Mason County WV along the Ohio river a man and a woman witnessed the incredible sight of a Living UFO. This couple started to head home from Huntington WV along the river, when they had an unbelievable encounter with a large Manta Ray shaped UFO. This "Animal" was doing figure eights over the road and river. This behavior has been seen in many open ocean filter feeders including Manta Rays. They watched this creature for some time before it gently flapped its wings and started to head back up towards the sky. A day later a woman and her daughter in Randolph County (a short distance from the first encounter) had seen a similar "animal" flying over the car. Two interesting facts with these encounters are that both parties had seen a similar creature and had no fear but more of a sense of wonder and that both encounters had been reported before either had been leaked to the press so neither party knew about the other encounters till after they were both reported.

Speculative Biology Analyses:

This is one of the more famous encounters with the "Manta Ray" body type. The animal or animals being seen in these reports are car sized and seem very slow and graceful. The behaviors seem to mimic that of open ocean filter feeders like mentioned above. This motion is used to collect small food particles or prey items (insects, evaporation algae and plankton and even small animals). This motion creates a small vortex that helps pull in its small food towards the mouth area. This style of feeding is very effective and helps large animals get larger. This account serves to help understand the biology of these sky creatures by comparing it to animals they are sharing similar ecological niches with regardless of their very different environments.

Title: Kansas Manta Ray

Location: Central Kansas

Date: Between 2001 and 2002

Witnesses: Wishes to Remain Anonymous

Encounter:

This encounter comes directly from the eyewitness; the account is unedited and proceeds as follows:

"Back in around 01-02' timeframe, a friend and I was laying on the hood of a car talking and looking up at the night sky. We were talking about the stars and personal relationship type stuff. It's a clear night in Kansas and the stars are very visible and the moon has a good amount of light in the sky. As we are talking, we both pause for about 30 seconds as we see something very large flying gracefully just above the tree line. It was flying slow and its "wings" moved like that of a manta ray. The front of its "head" was slightly longer than what a manta rays head is like, but it moved slowly as well. It made absolutely no noise, and produced no wind that should have disturbed the trees only about 30 ft below it. You could see through it the same as looking through clear gelatin and make out the stars on the other side of it. I thought maybe I was tired and my mind had imagined it. I feel tears welling up at my eyes due to the awe inspiring event that happened but brushed the thought aside. I looked to my friend sitting next to me and she is bawling and asks “Did you see that?”. I asked her if she saw something flying too. I tried to be as vague as possible to make sure we saw the same thing and it was. Mind you this was in Kansas nowhere near a body of water. Closest river was about 50 miles away. Its size was close to an airliner, maybe a little smaller due to depth and perception. Completely silent though. And the movements was very creature like, not that of something man made.”

Speculative Biology Analyses:

The creature in the account seemed to display many of the same characteristics of the manta ray class of upper atmospheric life from. It is seen flapping its “wings” very slowly but not affecting the trees below it, suggesting that it was not great, so it had little effect on the surrounding structures and that the flapping is most likely just used to steer its body rather than propel it. The next common denominator is that it's almost completely transparent, with many other accounts witnesses say they can see through them like they're looking at a translucent jellyfish.

This encounter was also at night, similar to several other accounts in this category. This could be due to them being closer to the surface of the earth at night or just being more easily seen at this time.

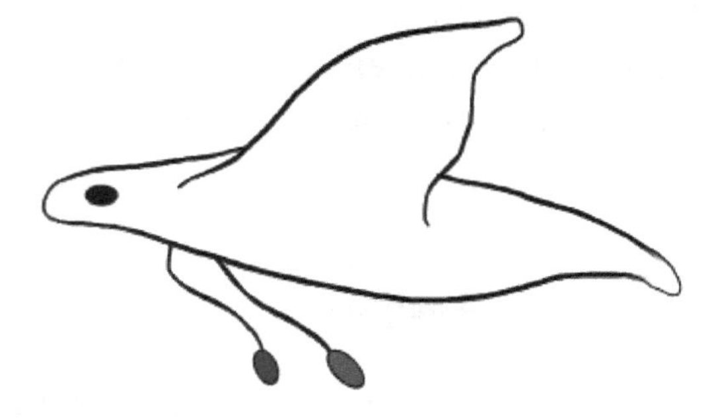

Art Credit: Mr. E

Title: Manta Ray Eating Dogs Over Chile

Location: North Chile and California

Date: First Sighting Sept. 29th, 2013

Witnesses: Unidentified

Encounter:

This creature was seen all over northern Chile, specifically in the areas of Bustamante Park in Chile, and later, even California. This entity has been observed to have the appearance of a Manta Ray, mixed with a man. Having four appendages with the front pair being giant leather wings, like that of an ocean ray, and the back pair trailing behind being like humanoid legs tails. This creature also has a small head-like structure in front of its wings. From some angles, it has the appearance of a flying human, but others angles look nothing like that of the human form. Most size reports put the body length at around six feet long and thinly built. This type of flying Manta Ray has been known to grab and eat dogs near Bustamante Park, swooping down and eating on their carcasses or carrying them away completely. The park mentioned above has had dozens of encounters with this aggressive monster, but the first documented sighting was

September 29th, 2013, when a local young man claimed to see the creature flying in and out of the tree canopy. There is a report from California that fits a very similar M.O. but there is not enough evidence to substantiate these reports. There are no named witnesses in these encounters due to locals being afraid of ridicule and work/life destruction. This string of sightings is still very active on America's coastline along the Pacific Ocean.

Speculative Biology Analyses:

The creatures in this group of sightings, while much more aggressive and smaller, physically resemble the classic Manta Ray UFO shape. As far as a biological reason for these creatures to exist, there are two groups of thought on these dog eaters. Firstly, they could be a sub-species or distant cousin. They could be a much smaller, and more predatory species, still using the same biological flying mechanism. Their smaller size could mean they're not going as high in the atmospheric column and could explain their dietary preference. Another thought is that these are the young of the gigantic Manta rays. Their increased aggression could be explained because they may not have the size safety of the adult forms and lashing out may make them not worth predator's time. The change in diet may be more common than one might think. Titanosaurs, the largest land animals on the planet, were heavy weight herbivores. As adults, most species would not find them as prey items unless severely injured, and they mostly kept to their own unless provoked. Their offspring, on the other hand, have some evidence to suggest after hatching, they would be opportunistic, acting as carnivores, feeding on, or even engaging in hunting other animals. The theory behind this is it would be easy to grow out of the food chain faster by taking advantage of all available calories. This could be what's happening with the manta rays. The Young Manta Rays are taking advantage of heavily nutritious animals until they're big enough to safely travel back up into the higher levels of atmosphere where the adults live. As for the area they're seen, this could be a breeding area or a nursery area for these upper atmospheric giants. We know so little about these magnificent creatures, this may be a future area of importance for the species.

Manta-Man as depicted in the novel, *Winslow Hoffner's High-Sailing Adventures,* written and illustrated by Michael Thompson. Image used with permission. Copyright © Michael Thompson.

Title: Triangle Gives City a Light Show

Location: Bogota, Colombia

Date: March 28th, 2011

Witnesses: Few Hundred People of the City of Bogota

Encounter:

In the early year of 2011, on a warm southern hemisphere summer night in Bogota, Colombia, an incredible display of light would fly over the city. This encounter was recorded by about half a dozen witnesses, so you can see this event for yourself. This city-wide encounter starts with three large, white lights coming over the city. They appear to be connected to each other in some way, but nothing can be seen in between the lights. As the lights move over the city in unison, they appear to be a part of a larger craft that is gently "flapping" as if attached to a pliable structure. The craft moves over the city and starts to head for the mountain range on the far side of the populated area and into an oncoming thunderstorm. The craft then disappears into the storm system.

Speculative Biology Analyses:

The craft in this encounter fits the description of a lot of the classic black triangle UFO, except for the gentle flapping witnessed in the apparent soft structures of the "wings" of the craft. The encounter precedes a thunderstorm, which is common with our organic UFO sightings. They seem to either be knocked down from their upper atmosphere habitat, so they are at the storm's mercy, or they are using the high-pressure front as an "elevator" lift to get back up to the stratosphere. As far as these large creatures being at the mercy of storm systems, we see the same thing with the largest beings on our planet, whales. Humpback whales can become severely discombobulated in severe storms, which can result in them being lost, in too shallow of water, drowning or even being beached. As far as these animals using these storms as lifts back to their preferred habitat, we see similar behaviors in birds like the California Condor which uses the thermal pressure wave in front of thunderstorm systems to rise above the weather system and use the air streams for easier long-distance travel. The entity in this event could be utilizing either of these transportation options. There is just not enough evidence in the videos and eyewitness statements to know for certain.

Title: Manta Rays in the Bayou

Location: West Monroe, Louisiana

Date: September 2022

Witnesses: Two Relatives

Encounter: In late September, a group of relatives in Louisiana were stargazing, trying to witness Jupiter and its moons. This was an astrological event that made it visible for the first time in several years. The group of people had low-grade binoculars that they used to search the sky for the cosmic event. Having seen little more than blurry lights, they stopped using binoculars and started using the naked eye. After several minutes searching for Jupiter with the naked eye, they found it. After a few minutes of observing, they find the Big Dipper. As they are watching the Big Dipper, they see something swishing across the sky. It was a gigantic, almost see-through, Manta Ray-shaped animal. It was like it was flapping its wings, almost as if it was swimming. They watched the animal for over 2 minutes, as it was zigzagging across the sky, until all too suddenly, it was gone without a trace. They expressed that they felt it was 100% an animal and they had a sense of awe while witnessing this creature.

Speculative Biology Analyses: The large animal with the zigzagging pattern, in the open night sky, is reminiscent of both the feeding patterns of large ocean animals, and our theoretical feeding pattern of these atmospheric creatures. The sudden disappearance at the end could be caused by several things, from the animal employing some kind of camouflage we haven't seen very often, to the witness just having a hard time tracking an almost translucent animal in a clear night sky. One other reason may be that these animals could be able to add incredible bursts of speed for very short distances as a way to get away from predators, similar to squid and octopus shooting jets of water.

Title: Translucent Manta Ray Over Florida

Location: Miami, Florida

Date: September 19th, 2014

Witness: Unidentified

Encounter:

Witness claims to see a huge Manta Ray shaped object flying over his residence, moving from the northwest to the southeast. It was transparent enough that the stars could be seen through the creature. However, the border of this creature was visible, having a white or gray tinge. The creature moved silently and slowly, and the witness estimated it to be about 500 feet from end to end. The witness felt awestruck and watched the creature until it was completely out of view.

Speculative Biology Analyses:

The creature in this encounter fits the description of the Manta Ray like entities perfectly. The witness felt no threat from the creature as it gently went by. This observation is one of the largest of the Manta Ray forms we see, at an estimated five hundred feet long, rivaling the jellyfish-like creatures as the biggest creatures in our skies. The noted flight direction suggests the creature was coming from the interior of Florida, heading to the open ocean. This is important because the ocean has thermal currents that would make it much easier for a creature of this size to reduce energy to fly long distances or to even return to the upper atmosphere.

Art Credit: Mr. E

Title: Manta Rays in the Desert

Location: Provo, Utah

Date: The Last Several Decades

Witnesses: Dozens of Reports

Encounter: In Provo, Utah, there have been dozens of reports over several decades of Manta-Ray shaped objects or creatures flying over and in the town. A lot of these creatures are around the 15-foot mark, and adhere to being very dark in color, from blacks to blues, with a lighter colored underbelly. These animals had even been witnessed hanging onto buildings, like the large church tower, until being realized, when they hop off and fly away at great speeds. Whenever these creatures are active in the midtown area, they do not appear to be staying for very long with most clusters of sightings only lasting a couple days.

Speculative Biology Analyses: Provo provides us with an interesting look at what may be the middle stage of this creature's life cycle. We've talked about the coastline of the Pacific Americas having many of the small Ray-shaped creatures, and as they're working their way inland to the larger mounds for the uplift to the upper atmosphere, they're growing in size. Unlike the titanic adults, these creatures still have their protective layers of skin, which could be why they look darker in color, with their keratin outside or thicker skin. Provo may be one of the last places on the ground for water or other resources before flying to the great deserts, mountain ranges, and plains. If these really are the juveniles of the titanic see-through manta rays, this area could provide great insight to the complete lifecycle of these mighty animals.

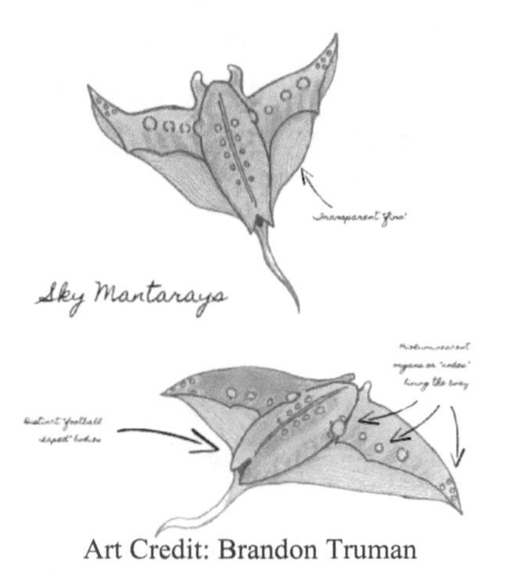

Art Credit: Brandon Truman

Sky Squids

Speculative Biology:

Sky squids are very similar to our ocean's blanket octopus, in which this form is very pliable and can change rapidly but do have a couple of forms they favor. The first form is a triangular like shape, which resembles a kite without the streamers. The second form is very Octopoda in structure. By this, I mean a more classic squid shape, with a long tube-like head and tentacles below the base of the "head". The final form is commonly seen as the kite-like form with two long, flowing "tails", which can be seen at half a mile long. In this form, they are often seen in pairs indicating that this may be a breeding display. Their flight abilities are partially similar to the Manta Rays, likely using a swim bladder like organ to maintain altitude and using their "fins" to steer. They also seem to be able to furl in their "fins" and "tails" and become ovoid in shape to gain speed when necessary. Their possible niche in the upper atmospheric food chain is that of a medium level predator, eating some smaller creatures on this list but still having to avoid others that may prey upon them. Their body plans witnessed are really two-fold. First is that their bodies are tubular in shape with some tentacle-like structures hanging off. These are often transparent, or mostly transparent, so most witnesses accidentally see them due to camouflage failure. The next most common type is when they're on full display, their large, flat, kite-like bodies and two long tail-like structures trailing behind them. There are most often two creatures that are intertwining each other, which is very similar to many mating displays in nature.

Art Credit: Michael Thompson

Sightings

Title: Belarussian Sky Squid

Location: Belarus

Date: 1985, 1999

Witnesses: Several Unnamed Airplane Pilots

Encounter:

In 1985, a pilot enroute to Tallinn, witnessed a cigar-shaped UFO with trailing tentacles flying overhead. The pilots aboard the craft said the creature was blue and red in color, brighter than the clouds around it, with a very smooth, asymmetrical body that was slightly transparent. It was moving in and out in a northeastern direction, climbing to the upper atmosphere. Both pilots claimed that the craft had a slight glow or fluorescence to it. The object made no noise and was moving at an incredible speed until it seemed to vanish at a very high altitude.

In 1999 a very similar creature was seen by two pilots flying a routine mission. Almost matching the exact description of the first creature, when later shown pictures, they said it looked like a gigantic glass squid. This creature again had the blue and red lights with a slightly transparent body as they watched it go higher and higher in the atmosphere. The pilots say there was no aircraft that could go as high as the object or creature during the time. That is why the appearance and altitude were both noted with this strange sighting.

Speculative Biology Analyses: With both Belarus sightings, two very interesting things were noted; the large transparent body with tentacles trailing behind it, and the speed and the direction the creature was heading. The lights moving around the body are similarly noted in ocean cephalopods and can be used as a defense mechanism. This, combined with the speed at which the creature was returning to the upper atmosphere, suggests they may be using this as a similar survival tactic. The planes could have scared or threatened these creatures, causing them to flash their lights as a threat display as they retreated to where they felt safe in the upper atmosphere.

Title: Blanket Octopus of the Troposphere

Location: Somewhere Between Phoenix and Portland

Date: 2019

Witnesses: Crew and Passengers of a Commercial Airliner

Encounter: A commercial airliner, flying from Phoenix to Portland, witnessed what the pilot thought was a small plane on fire, with two large smoke trails behind it. The crew was able to catch a video, in which the creature makes an upward turn to be parallel with the plane. You can see that the creature is black, but almost transparent with two large "tails" trailing behind it.

Speculative Biology Analyses: In the video, you can see in a very short time, the creature goes from almost downward angle to parallel with the plane in almost matching its speed. The creature in the video is very similar to the ocean creature known as a blanket octopus. With their two long trailing tentacles, they shape their bodies to match predators and prey for both defense and hunting. It is unknown what the intentions were with changing its direction to match the plane, whether it thought it was a rival, food, or threat, but with the speed it changed direction, it definitely seemed to take notice of the passenger plane.

Sky Siphonophore (Sky Snakes)

Speculative Biology:

This group can have very confusing biology to look at, with eyes in their mouths, multiple undulating pairs of fins, partially transparent bodies, and very hard to define heads. This group's superficial appearance does, however, have a matching earthly group of animals called Siphonophores. These amazing animals are very worm-like in appearance but have segmented body styles that are similar to some worms. Siphonophores are colonial species, which means the mass of their bodies are a buildup of thousands of individual animals that are extremely specialized. Some animals act as the fins, some aid in digestion, and some steer the body. These groups of animals can look just like the famous sky snakes. As far as Sky Siphonophores go, they have been seen almost "swimming" through the sky on multiple pairs of fins. Their abilities of flight and locomotion are likely due to their segmented body having a gas filled sac in each body segment, using their quick moving fins to steer and get up to incredible speeds. People have observed them having what they perceive as extremely large tooth-like structures in their mouths. They have been observed being over two hundred feet long. Often, when they're seen close to the ground, they seem to be in very poor health condition. Unlike some of the other types of sky creatures, they may not be able to return to the upper layers of the atmosphere. As far as their niche in the upper atmosphere is concerned, these are our sharks. They seem to be built tougher than other organisms that we have observed, including ridged heads with large teeth, streamline bodies for speed, and their overall size. Their body would allow them to hunt most of the other creatures in their domain besides the biggest Manta Rays. They have been witnessed to move very fast and turn on a dime, not dissimilar to open ocean sharks, like Blue Sharks and Makos.

Sightings:

Title: Twister Worm

Location: Oklahoma City, Oklahoma

Date: May 3rd, 1999

Witnesses: Danny Romero and Other Local Residents

Encounter:

Danny was a resident of Oklahoma City, who was an avid storm chaser. One evening, he experienced something he's never experienced before. May 3rd, 1999, a tornado touched down with the fastest speeds recorded on Earth. Danny watched from his porch as the tornado approached, and it was at this time he started recording. He witnessed, during a lightning flash, what appeared to be a long piece of corrugated tubing alongside the tornado. Upon further inspection, it appeared to be a living creature, with a long translucent worm-like body, and a wide head like a hammerhead shark. He speculated the length of the creature to be at least 150 feet long. It appeared to be swimming in and out of the tornado, accompanied by smaller similar creatures. The creatures stayed in the storm system until completely out of eyeshot of the witness. This encounter bothered him so much that he gave up storm chasing completely.

Speculative Biology Analyses:

(Possible hoax) The eyewitness is under the impression that the creature is actively swimming through the tornado, but it seems more likely that the creature was caught in the storm system, ending up in the tornado where it appears to be struggling. The smaller "creatures" witnessed are most likely strips of flash coming off the large creature due to the whipping winds and the debris in the storm. As far as the creature's unique anatomy, there are a few oceanic animals that fit some of these oddities too perfectly. The hammerhead shape of the head seems to be, more likely, long jaws that are locked wide open like that of the Bobbitt worm. The Bobbitt worm is a long segment body and appears to have a hammerhead, but in reality, it's the animals' jaws locked open with a hair pin trigger to slam shut on prey. The Twister worm seems to fit a similar biological design to the Bobbitt worms' jaws giving it that distinct look. The frightening appearance of this creature bothered the witness so much he gave up his great passion, so the creature must have been quite the sight to see.

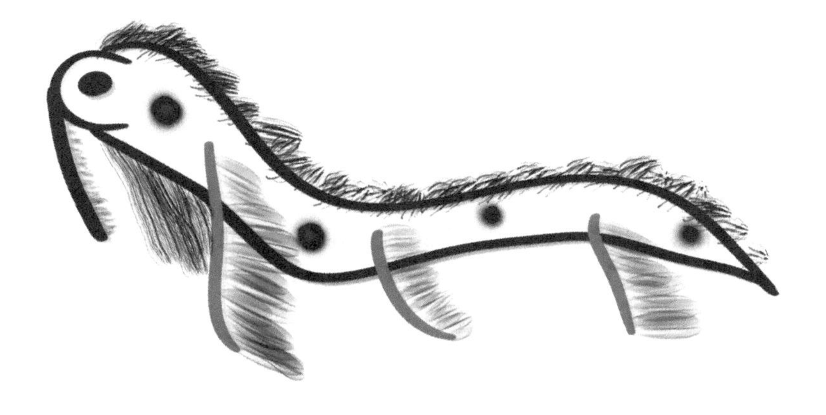

Art Credit: Mr. E

Title: Sky Serpent Over New Delhi

Location: New Delhi, India

Dates: May 17th, 2008 / April 28th, 2014 / October 30th, 2015

Witnesses: Multiple Witnesses

Encounter:

The first reported sighting over New Delhi occurred on May 17th, 2008, when a resident was flying from New Delhi to Guwahati. He took out his phone to take aerial photographs. It was at this moment he captured what appeared to be a coiled, glowing snake.

The second sighting occurred on April 28th, 2014, above the city of New Delhi, where a yellow, glowing ring appeared, floating in the sky. One witness claimed, "It had a bright red dazzle around the top half of its body."

The final sightings to date took place from October 27th - 30th, 2015. Air traffic controllers reported to The Hindustan Times, Oct. 27th, a UFO that did not show up on radar equipment. Three days later, an Indian Air Force officer reported seeing three UFOs around the airport from the runway tower. This prompted a meeting to occur, which gave the Indian Air Force a shoot-down order. The three UFOs looked to be long, serpentine, glowing creatures that matched the previous sightings in the same area.

Speculative Biology Analyses:

This area has been a hotbed of sightings of these creatures for the last decade. Most of the types of creatures around the world are normally only seen with glowing bodies of one color, often white or yellow, but this area has been known to produce creatures with red light on their body as well. Due to the constant number of sightings, and the somewhat unique colors of bioluminescence, we may be led to believe that this is a breeding area. The number of creatures seen throughout this area, including multiple creatures at the same time, and the extra lights displayed, point to some kind of mating site or important area for reproduction. These creatures are also observed moving incredibly fast, with our own jets having trouble contacting them. This shows the capabilities of these creatures to move when they need to.

Title: Glowing Serpent Above the Desert

Location: Mojave Desert

Date: June 2019

Witnesses: YouTuber Cody Kennedy

Encounter:

The witness started following the UFO, which appeared to be stationary at first, but upon getting closer they realized it was moving quickly, covering a great amount of distance. The object was glowing a bright pulsating light, and the front appeared to be stationary while the backend looked to be flapping in the wind. The front end stayed a constant glow, while the lower half pulsated light down "the body".

Speculative Biology Analyses:

The creature in the video is hard to get a size estimation on, due to being alone in the night sky, but appears to be very large. It is also moving incredibly fast, with the witness having to get into a vehicle to keep up. With the front of the animal remaining very smooth in its movements while the tail end of the creature flapped or wagged wildly, it seemed to either be using the tail for communication, threat display, or a mating display, which is seen in animals like the Axolotl, who use tail wagging for all three. Deep-sea fish use light displays for the same thing. If it's for communication, the pulsating light pattern could be a complex message to other members of its species about location of food, or threats. If it’s a threat display, the wild movement and rapid flashing of lights could be used to confuse or disorient a predator, which we see all over the natural world. As for a mating display, we see this in open water squids, with males with the brightest light and the wildest movements being the healthiest, and most breed able males.

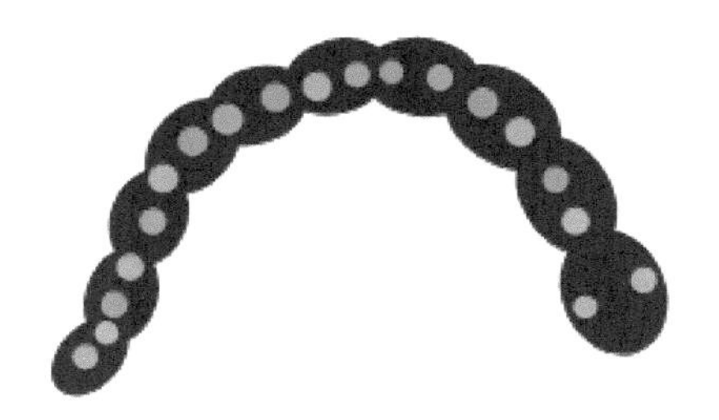

Art Credit: Mr. E

Title: Black Sky Snake

Location: Canada

Date: 2013

Witnesses: Unknown

Encounter:

Over an area of Canada, there is video of a very large, serpentine, black object in the sky. This object appears to be undulating very slowly, reminiscent of a deep-sea worm swimming through the sky. Without context, it's hard to tell how long this object is, but appears to be over a quarter mile in length. The creature in the video was observed for well over 3 minutes, but due to its slow speed, was probably visible for a long duration of time.

Speculative Biology Analyses:

The video shows a massive creature. It is impossible to know for certain how large the animal is, but it was estimated to be between 6 and 8 hundred feet long and approximately 20 feet wide. An absolutely massive creature! The creature is undulating very slowly in a movement from side to side, like a free-swimming ocean worm, but you can also make out smaller movements on each body segment, suggesting smaller pairs of fins that may help with locomotion and steering. The appearance of slow speed is misleading due to the distance the creature is from the camera. It is most likely moving quite fast. This is one of the bigger members of this group of living UFO's.

Title: Rainbow Serpent Over the UK

Location: Blankney Barff, Lincolnshire, UK

Date: 11/08/2015

Witnesses: One Unnamed Eyewitness

Encounter: Around 10:30 PM, the eyewitness was staring at the sky for some stress relief and relaxation. At this time, they were made aware of the huge serpentine-like string of lights moving across the sky. The lights were red, green and blue. The body of the animal was very long. It is impossible to tell from the distance of the photo provided on the web site, but you can see how huge it appeared to be in the sky. As the creature moved across the sky, it was undulating back and forth, so when the witness took the picture, it was in a U-shaped.

Speculative Biology Analyses: The dramatic light show put on by the creature is very reminiscent of scenes from the deep ocean itself. If there was video proof that this creature was in the upper atmosphere, it would look almost exactly like the bioluminescent Siphonophore in which the category is named after. This little deep-sea group of organisms often have long serpentine bodies with multiple sets of hairs or swimmerets and are often very bioluminescent. We still don't know fully why these creatures glow the way they do, but some of the options are to attract prey, to scare away predators and, most likely, communication or sexual display for a mate.

Title: Black Worm Over New York

Location: Clayville, New York

Date: 2019

Witnesses: Unidentified

Encounter:

In this video encounter, another large, black, worm-like UFO is seen slowly moving from a vertical position to a horizontal position. The creature, at some points, seems to be free-falling and at other points has controlled flight. Once again, there is no size reference in the sky, so it is hard to estimate, but it does appear to be quite large.

Speculative Biology Analyses: With the video evidence provided, it's very hard to tell whether this is a biological creature or some type of extremely large tube falling from the upper atmosphere. If this is an organic creature, it is safe to say that it may be falling from the stratosphere or even higher up. Several times, it looks like it is stopping its flight or slowing down, thus, it doesn't seem to be in full freefall but appears to have trouble staying in the sky. This could be due to a predator strike, causing the animal to struggle to stay aloft.

Title: The Crawfordsville Monster

Location: Crawfordsville, Indiana

Date: 1891

Witnesses: Almost an Entire Town

Encounter: The town of Crawfordsville, Indiana was in fear when a very strange creature came flying through their skies. The town would describe this creature as being roughly 20 feet long and 8 feet wide, quoting as saying the "giant flapping thing with a huge flaming red eye". It was said to have a very eel-like body with large feathery protrusions running down the side, with a mouth that split into many parts. One witness claimed the eye was in the center of the mouth. Many of the witnesses reported that they felt the creature was in great pain, describing everything from writhing and squirming, to claiming that the creature was wheezing in a sickly painful manner. Rev. G. W. Switzer and his wife saw the animal and wrote of their account, generally agreeing with the public's description of the creature. He and his wife also said that they thought the creature was squirming in agony and sounded like it was making a wheezing noise as if it was near death. He also estimated the height of the creature to be about 300 feet off the ground, claiming that it seemed like it was having trouble maintaining that height. Most of the eyewitness testimonies from this day claim that the creature most likely crashed and died in a field located just outside of town, although no substantial evidence can be found of this.

Speculative Biology Analyses: This is another perfect example of what may be convergent evolution. This creature sounds extremely similar to some deep-sea worms and even some famous shoreline worms like the Bobbitt worm. A lot of these oceanic worms, like the Bobbitt worm, have feathery appendages that they use for swimming, digging, or crawling. These creatures also have some of the weirdest jaws in the animal kingdom, sometimes appearing nightly with hammerheads like that of the famous shark, but it's really their jaws. The jaws appear to be in the hammerhead shape but are wide open and snap shut to secure prey. The Crawfordsville monster has a lot of the similarities to oceanic worms, so it may be using these feathery appendages to steer itself to the upper atmosphere, or maybe even for propulsion, having gas filled cavities in its body to help maintain buoyancy in the atmosphere. The wheezing sound the town heard could be a multitude of things, from the creature's internal gas chambers leaking from a fight causing its own lose altitude and eventually die, or even the creature just being sick like that of some whales and the ocean, not being able to return back to the surface or for these creatures, the stratosphere, and eventually crashing and dying. The giant red 'eye' could be the esophagus or a primitive tube mouth like that of modern tarantulas and scorpions. This is one of more fascinating cases we have seen due to the sheer number of eyewitnesses, with the date being before almost anything in the sky was deemed a balloon, the detail of the creature agreed upon by the town, and almost everyone that saw it agreeing they felt it was ill.

Atmospheric Amoebas

Speculative Biology:

This group is something we know very little about and by the time the second book comes out, hopefully will be explored more deeply. These creatures are best described as floating, amorphous blobs or living clouds. These creatures are often huge with no clear, defined shape, often mistaken for fog or clouds. They do appear to be carnivorous, with the raining blood events and the raining bone events attributed to these creatures. We think they may be much more scavenger oriented and probably only prey upon creatures too slow to move out of the way. Some of the places in the world have long-term histories with living fogs, often warning of how to avoid them. We have several theories about their proposed biology. One of the first ones is that they may be able to create a light layer of vaper around their bodies to help break up their shape to both avoid predation and to help move closer to unsuspecting prey sources. Another thought is that these creatures may be tightly packed schools of smaller atmospheric creatures, feasting upon whatever creatures enter the swarm, much like coastal sea fleas and the dramatic retelling of how piranha behave. If that is the case, these creatures may be the planet's alternate cleanup crew working through everything from dead flesh from the upper atmosphere to rotting plant matter on the surface and we would only notice them when they are condensed in tightly packed clusters. The amoeba-like giant creatures were some of the first organic UFO's ever reported in this field of study. It would make sense if these animals operated as giant hot air balloons.

Art Credit: Mr. E

Sightings:

Title: Raining Bones

Location: Carroll Parish, Louisiana

Date: June 21st, 1872

Witnesses: Local Residents Contacted Paper

Sources: Library of Congress

Encounter:

The following event was documented in newspapers of the time for its incredible bizarreness. This encounter takes place in a small parish in Louisiana. When a heavy storm system was passing through, a large black cloud was passing low over the town, dumping a large amount of fish bones and what appear to be gar scales all over the town. This event may shed more light on their biological niche in the lower atmosphere environment.

The article from the Nashville Union and American – July 11, 1872 – Tennessee, stated the following bizarre occurrence.

"If the statements of some of the residents of Louisiana are to be credited, Dame Nature has recently been playing strange pranks in that part of the country. A writer to the New York *Journal of Commerce*, whose veracity and good standing is vouched for by the editor of that paper, gives the following particulars of a strange phenomenon that occurred in Carroll Parish last month:

He says that a heavy storm visited that parish some days previous to the date of writing, the 21st ult., and during the storm fish bones fell to the ground by the million. These bones seemed to come from the exceedingly large black cloud that was passing at the time. The shower of bones was attended by a heavy fall of rain.

The correspondent says that the bones rattled on the roof of his house like hailstones. This strange phenomenon extended over a belt of country ten miles in width by many miles in length. Accompanying the letter were seven of the bones varying from one inch to two inches and one-sixteenth in length, from seven-sixteenths of an inch to twelve and half sixteenths of an inch in breadth from one inch to one inch and nine-sixteenths in length, and from one and a half to three-sixteenths of an inch in thickness. They are of an irregular diamond shape. One side of the bones is nearly flat, having on the under side, which is worn smooth, three small apertures, as if veins or tendons had passed through them. These specimens have been shown to experienced fishermen, and also to learned ichthyologists, but they are not able to ascertain to what particular kind of fish the bones belonged. They all agree, however, in the opinion that they are veritable fish bones.

Several theories have been advanced in explanation of this strange phenomenon. It is generally conceded, however, that the bones must have passed through the air for hundreds, and perhaps thousands of miles. The inhabitants of the parish believe that they were brought by a water-spout or a whirlwind from the western coast of Mexico or Lower California, across the continent, as the wind was blowing at the time violently from the Southeast.

We have heard of it raining cats and dogs, but fish-bone showers are something altogether unprecedented." This is the article in its entirety from the Nashville Union and American.

Speculative Biology Analyses:

These creatures are known to have a striking resemblance to fogs or clouds, like some of the jellyfish types. From personal experience, in Louisiana after heavy storms, fish end up in oxbow lakes and dams, specifically large Gar. Gar scales are hard and bony, and diamond in shape which fits the description of the scales that fell out of the cloud. As far as behaviors presented, if this dark cloud was indeed one of these Atmospheric Amoebas, it could have taken advantage of the large groups of trapped fish or dead fish in the local area. After filling itself with fish and starting the digestion process, it could have been regurgitating the parts it could not digest, like the bones and scales. For scavengers, it's very common to overeat due to the uncertainty of another meal. Hyenas and vultures will eat on a carcass until they throw up to continue eating.

Title: An Amorphous Cloud

Location: Unconfirmed Canadian Sighting

Date: 2021

Witnesses: Unknown Woman

Encounter: In this woman's video recording, is what appears to be a cloud flying very low over the highway. The edge of the cloud and the overall density seem to be much thicker than typically expected. Several times during the video it appears to be almost pulsating like a jellyfish or some other oceanic gelatinous creatures. The video is very short, so any further details are very hard to get. The cloud almost appears to have an orange-colored center to it.

Speculative Biology Analyses: If this is indeed a creature, it seems to pulsate as if trying to gain lift again. It very much appears to be, in appearance, almost like a water balloon about to burst. You Can see a clearly defined edge of the cloud, which is very atypical for a normal cloud. As it pulsates, you can see it appears it's almost made of jelly and several times in the video it appears to have an orange glow emanating from the inside. A lot of these creatures are seen right before, during, and immediately after storms and from her video you can see a storm starting to form in the background. This creature may have been feeding nearby and is either trying to avoid the storm or got knocked out of its feeding range by the storm.

Art Credit: Mr. E

Jellyfish-Like

Speculative Biology:

This group is almost completely identical to their ocean counterparts in looks and behavior, but there are major differences in size. These creatures have been witnessed to be of unimaginable size; four hundred feet wide and six hundred feet long. They have also been seen to be incredibly slow moving, almost drifting around at the will of the wind current. They have also been witnessed to display incredible bioluminescent light shows. Witnesses who have firsthand accounts compare it to watching a blimp covered in LEDs that flash very complex patterns of white, blue, red, green, and even yellow lights. The resemblance to actual Jellyfish is so uncanny that it leads one to believe they may be more related than first realized. In some of the more recent upper atmospheric life studies, a very small Jellyfish relative has been found, which makes the thought of this group being closely related to the Jellyfish family an even greater possibility of being factual. In behaviors, they have been seen to drift around very slowly with a gentle pulsing of their "bell". Their ability to stay airborne is likely due to their large center chamber being filled with gas to maintain altitude and using the gentle pulsing to steer. They also are followed by hundreds of tendrils flowing from the center of their body, trailing along behind them at the mercy of the wind. Their niche seems to be another giant filter feeder. We have giant Jellyfish in the ocean that occupy the same niche as the Lion's Mane Jellyfish. These creatures prevent becoming prey items from faster moving predators by being very poor in nutrition and being armed with very potent venom.

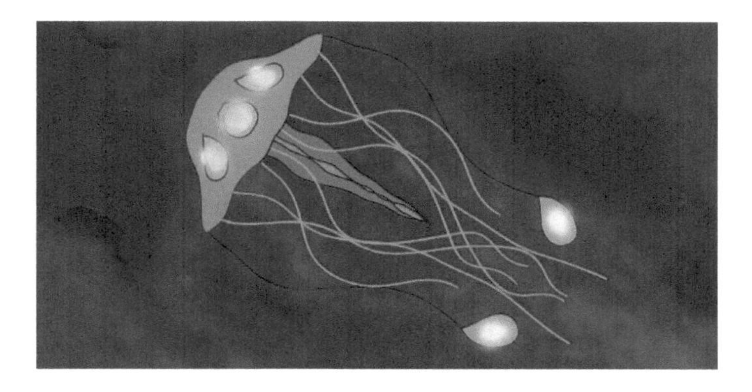

Art Credit: Mr. E

Sightings:

Title: The Dutch Jellyfish

Location: Netherlands

Date: June 1st, 2015

Witnesses: Harry Perton

Encounter: A man by the name of Harry Perton ventured out after a storm had hit the mountains nearby, to take pictures of the beautiful sunset. He did not realize until he got home that he had captured a photo of what appears to be a giant, green colored, jellyfish shaped UFO. Though he did not see the UFO during the time of the photo shoot, he did claim he saw a large flash in the sky above them, causing him to nearly drop a camera lens. He dismissed this as a lightning strike until he saw the photo.

Speculative Biology Analyses: If this photo does, in fact, show an atmospheric jellyfish, it doesn't share a lot of anatomy for its oceanic name's sake. The creature lacks a bell and what appear to be tendrils. This could be some other atmospheric form, but from the single photo, it's almost impossible to tell.

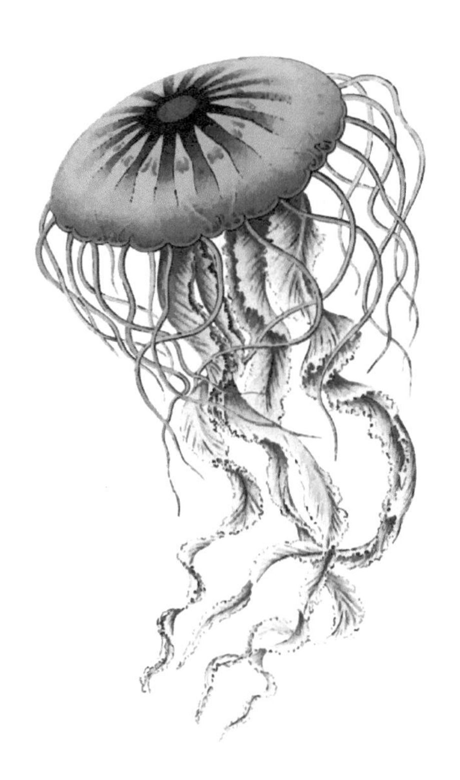

Title: Qing Xian Jellyfish

Location: China

Date: October 19, 1998

Witnesses: Several Chinese Air Force Pilots and 140 Ground Officials

Encounter:

On October 19, 1998, four military radar stations in Hebei province, China, reported the presence of an unidentified object hovering slowly over a military flight training school. One story determined that the object was not a military or civilian aircraft. Colonel Li, who was the base commander, ordered six fighter jets to intersect the object. At least 140 ground personnel witnessed the object hovering over the base. Many of the observers at the base said the UFO first appeared to be some kind of small star. It would grow larger and larger, but they weren't sure it was descending in altitude. As the object got closer, they all said that it was very jellyfish-like in structure, with the bottom having long, hanging tentacles that were covered in bright lights. The crew in the six fighter jets intercepted the object, radioing back to the tower that whatever they were seeing was straight out of science fiction, claiming it to be some kind of jellyfish-like monster. As they got about 4,000 meters from the object, it abruptly shot up, easily avoiding the fighter jets. The fighter pilots continued trying to engage the target and reported that it felt like it was toying with them repeatedly. Ground control that night issued an order that they discontinue pursuit of the object instead of having them continue trying to engage it by getting closer. The pilots followed the creature until roughly 40,000 feet, then stopped due to the jets running low on fuel. The whole time of the engagement, the pilots felt as though the creature could easily escape the fighter jets, but instead, chose to continually interact with them.

Speculative Biology Analyses: One of the most amazing parts of this encounter is that every eyewitness is military personnel. This occurred well before the UFO disclosure events, so these people had their reputation to lose. Another amazing thing with this is the variety of viewpoints this creature was seen from the pilot, ground personnel, radar and even interesting photographs. As far as the anatomy of this creature, it really speaks to what we think they may be. Giant gas filled bells with tentacles for feeding or communicating purposes. One of the other amazing parts of this encounter was that the pilots felt as if the creature was playing with them, as a dolphin does with a tugboat, knowing how much faster they are, but that they are not in any real danger. Continually dodging the jets, but allowing them to get closer, it could possibly have some kind of way of sensing the world around it, whether that be electromagnetic, or maybe even visually with eyes, although it's very hard to tell. With this amazing, documented behavior, it's safe to say that there may be more intelligence within these creatures than their oceanic counterparts. These creatures may be occupying the niche of giant filter feeders; they may be like modern whales and elephants, being too big for most predators to bother with, allowing them the potential to develop more social skills and intelligence than smaller creatures.

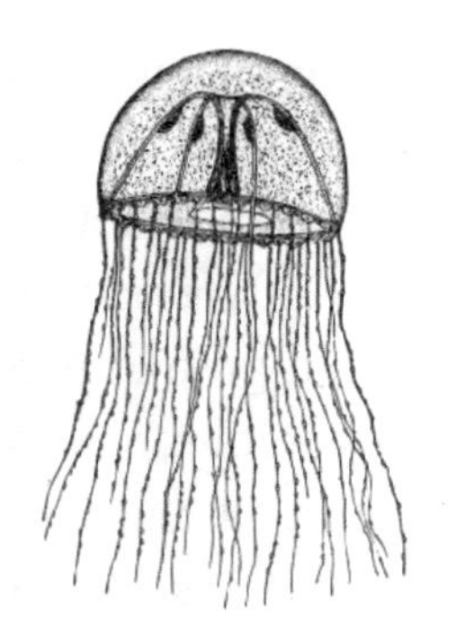

Title: NASA's Jellyfish Sighting

Location: Somewhere Over North America

Date: December 2012

Witnesses: NASA's Sky Cameras

Encounter: NASA sky cameras captured an interesting photo of what appears to be a large, green, jellyfish shaped entity flying across the sky somewhere over the North American continent. The figure appears to have either an extremely large thin bell, or what could be two large fan wings and was trailed by two long, thin appendages with multiple tiny other appendages. This photo was dumped into the NASA archives until the alien disclosure group went combing through them. Within the archives, they found this photo in January of the following year.

Speculative Biology Analyses: This photo shows what could be one of our extremely large atmospheric jellyfish. Not really knowing the camera's position and how high this creature is in the atmosphere, it's impossible to tell an exact size but it appears the animal is built quite large. The appearance of the large 'wings' could be the bell of this large jellyfish, filled with gas to allow the creature to stay aloft. Until more data is released from NASA about this photo, it is very hard to speculate anything more.

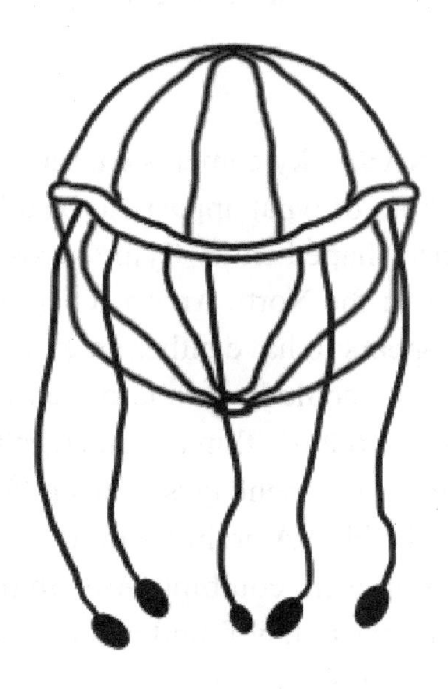

Art Credit: Mr. E

Title: Brazilian Tentacle Monster

Location: São Paulo, Brazil

Date: February 2021

Witnesses: Unnamed Filmer.

Encounter: In this video you can clearly see a living creature resembling that of the lion's mane jellyfish. The creature appears to be moving in an upward angle through the sky, gently pulsating its bell. The video shows hundreds of extremely long tentacles flowing behind it. One may think it's just a jellyfish filmed in the ocean, but it was shown several times throughout the day, and you can see clouds in the background.

Speculative Biology Analyses: The interesting thing about this video is that if this is indeed a real creature in the upper atmosphere, they are strikingly similar to the oceanic counterparts. We're still not sure if it is convergent evolution or if these animals share a common ancestor with oceanic jellyfish, then evolved into similar shapes in different environments. Something to analyze from this video are the extremely long tentacles. In jellyfish species, these extremely long flowing tentacles are normally used for passively catching prey items while they are drifting through the sea. What we speculate about the atmosphere variety is that they may be doing the same thing, slowly drifting into the atmosphere, snagging everything from detritus to other living creatures in the atmosphere. Like oceanic jellyfish, these creatures and their tentacles may be crucial habitat for the young of other atmospheric creatures, hiding amongst them for protection.

Living Silver Discs

Speculative Biology:

This category is reserved for what appears to be living disc UFO's. Most of the time these UFO's appear to have a metal exterior until upon closer examination, you realize it's some kind of hard shell that most of the time bisects in the middle of the craft. As far as we can tell from the sightings of these creatures, they appear to be very intelligent, both in hunting and, on rare occasions, even communicating with people. Their intelligence can also be attested by eyewitnesses who claim that these silvery creatures often play or toy with aircraft, from gently bopping biplanes to leaving fighter jets in the dust. These beings are often seen during extreme aerial maneuvers, from sudden stops and starts to 180° turns, going well above supersonic speeds. Their speculative role in the upper atmospheric environment as adults, they will often take the role of apex predator. We theorize they have some kind of mouth part but witnesses have also claimed they have strong maneuverable tentacles that come out of the sides of the shell. They may use their supersonic speeds, like falcons, to completely blitz their prey, taking them by surprise and hitting them hard enough to kill them. Not only are they fast, they also appear to be quite large, with some of the bigger specimens seen at well over 50 feet in diameter. These creatures also have an array of bioluminescence, with witnesses seeing all kinds of lights from super-sized spotlights to dozens, if not hundreds of small multicolored lights. From witness testimonies, it seems these creatures use these lights from everything from hunting for food to advanced communication. There have also been reported psychoactive effects from being too close to these creatures for long periods of time. Some of this group may also be eating radioactive materials, with lots of these creatures sighted around nuclear power plants, missile silos, metal waste dumps and heavily polluted areas. In the same vein as this, they also appear to have an affinity for fertilizers and other high nitrate foods.

Sightings:

Title: Communication With Lights

Location: Atlantic City, New Jersey

Date: December 2016

Witnesses: Two Anonymous Commercial Airline Pilots

Encounter: Two commercial airline pilots near Atlantic City, New Jersey noticed a very strange light in the sky. They took note; this was not an Astral body like Venus. The light was moving at an angle and keeping pace with the 600-mph commercial flight. Both pilots had several thousand hours of flight time between them. The pilot stated it was at approximately the 1:00 position when first seen, but it moved between the 12:00, 10:00, 11:00 and back to 12:00 position. Not only was it swaying back and forth in the forefront of the plane, it was also matching the incredible speeds with relative ease. They would also describe that it seemed like the object was getting larger and smaller. They weren't sure if this was due to a light anomaly from the object or if the object itself was pulsating. The pilots mentioned the light in that 1:00 position and shot straight to the 10:00 position before disappearing at astonishing speeds. During the time of the encounter, which lasted about 10 minutes, the copilot would flash the lights of the plane in a sequence. From what they could tell, every time, the UFO would match the same sequence. Both pilots were not sure of the intentions of the return sequencing, but assumed it was an attempt, in some way, of communication.

Speculative Biology Analyses: This is a very interesting case due to the fact we see that these creatures are capable of incredible speeds, yet still choose to interact with our craft. The entity here moves all around the front of the plane, keeping pace at over 600 miles per hour with relative ease, even surpassing it without ever audibly cracking the sound barrier. This type of behavior has been seen across the other categories as well, like dolphins playing with tugboats until they get bored and shoot off. Another fascinating detail is the reciprocated light show from the plane and the craft. On the basic level, this could be similar to how some researchers are able to use mimicking of squid lights in the deep ocean to bring squids closer. Many large deep-sea squids and other animals use complex light patterns in sequences to communicate. For them, communication is much more about color and patterns than it is verbal. So, the entity seen in this case could be actively trying to communicate or understand what the plane's intent was, being fascinated with another entity flashing lights back at it, and eventually getting bored and shooting off at several 1000 miles per hour.

Title: Crash Landing On The Mesa

Location: Nevada

Date: 1925

Witnesses: Don Wood Jr.

Encounter: Wood and three other men were flying Cessna JN-4 airplanes, commonly referred to as Jennings, through the deserts of Nevada. The group of men decided to land on the top of the mesa to enjoy lunch. During their time on the mesa, they saw a flash of light in the upper atmosphere. Not long after, a red colored desk descended slowly towards the mesa on a shaky trajectory. The disk was about 8 feet in diameter. After the disk crash landed, the group of men realized it was some kind of animal. It appeared to be breathing by raising its top half up and down, creating a six-inch opening along the whole rim of the top shell. The men compared it to a giant, metallic clam. As the men continued to investigate the creature, they realized it had a substantial chunk taken out of it. The creatures 'blood' coming out of the wound was a metallic looking ooze. As the men watched it, they stated that it made a wheezing sound and realized it was probably near death due to its injuries. After about 20 minutes, the animal started to glow with a pulsating red light. Gaining more intensity, the creature began trying to get back aloft. It would float in the air a little bit, but due to its injuries, would come crashing back down. As the men watched the creature the whole area was covered by a massive shadow. As they looked out, an even larger disk-shaped animal was seen floating down from the upper atmosphere. This creature would ignore the men as it settled over the other injured creature, latching onto it with sucker tipped tentacles. In a great burst of speed, the giant animal grabbed the little one, shot upwards into the sky, and vanished. The men were not sure if the larger creature was attempting to eat or save the little one.

Speculative Biology Analyses: For us, this is the case that started all of this. The unique biology witnessed by these men seemed too farfetched to be made up. The interesting silver shell that the creature apparently had, that could open up and expose the internals and the tentacles was unique and fascinating. Different reports of the same story said they could see several eyes, like that of sea snails. This creature appears to be made for high speed and faster maneuverability. The flashing the pilots had seen before landing may have been a school of the small creatures being hunted by the large one. After one of the small creatures escaped with a lethal injury, it made its way to the mesa, trying to recuperate. The metallic ooze coming out of the wound is very similar to some of the fluids inside of earthbound fungi. When the small creature began to try to take off again, and glow with red lights, it may have either been using a defense mechanism, trying to ward off the approach of the predator, or signaling for help from others of its kind. The men said the larger creature was similar, but not the same as the smaller creature which leads me to think this could be a close cousin that preys upon the smaller species that was looking for its lost meal, not caring about the men, grabbing its prey item with giant hooked tentacles like that of massive squids, and launching itself back up the higher levels of the atmosphere. It seems these men were witness to a rare predator and prey event with a close look at the anatomy of these amazing creatures.

Art Credit: Starcruiser Studios

Title: Aloha UFO

Location: Island of Hawaii

Date: 2022

Witnesses: Local Homeowner

Encounter: When a local homeowner came to his residence, he noticed a neighbor was staring at the sky. When he looked in the sky, he saw a large, white, metallic disk shaped "thing", that may have had small hair-like protrusions around the edge, shooting across the sky. He managed to take one photo of it. He then stated later that the U.S. military got involved with the craft.

Speculative Biology Analyses: We often talk about how these creatures may in fact be using ocean-like thermic currents for easier rides and may even be feeding off microorganisms that are similar to what they would find in oceanic environments.

Sky Jelly, Meat Showers and Spore Events

Speculative Biology:

This is a phenomenon that's been reported for thousands of years all around the world. Witnesses claim to see globs of blue, pink, or transparent slime falling from the sky. It's also documented to have chunks of unidentifiable "meat" raining over an area as large as 4 counties. Just as wild, black viscus snow has covered areas with no explanation. The slime has been coined as "star jelly". Most star jelly events are likely one of two main causes: reproduction, or death of a soft bodied sky creature. As far as reproductive byproducts, this substance has been seen in congruity with UFO's and may be their sex cells that would later form more of their kind or is what's left over after mating occurs. Animals like slugs and spittlebugs leave behind large amounts of slime and foamy mess with no sign of the animal or eggs after mating. As far as a deceased sky creature, both sky jelly and meat showers could be explained by this. If a very large gelatinous creature dies high in the atmosphere and starts to fall to earth, their corpse would start to be ripped apart when hitting terminal velocity. The body would be spread across a small area on the surface of the earth. This would also explain why in some cases, people get extremely sick after touching the pieces. Some body parts, like jellyfish stingers, can still sting for days after death. This could also be caused by upper atmospheric predation events, where some creature is hunting another and is ripping it apart. The scraps of the meal are what's hitting the ground. The black snow is likely a spore event. People who encounter this "snow" often report a mold-like smell around the site and have even reported respiratory problems very similar to encountering heavy constructions of black mold. The phenomena of black snow could be the reproductive strategies of aerial fungal UFO's and is likely very toxic.

Sightings:

Title: The Kentucky Meat Shower

Location: Bath County, Kentucky (but several other nearby counties reported the same thing).

Date: March 3, 1876

Witnesses: Several Hundred Eyewitnesses, Most Famously Mrs. Crouch

Encounter: On March 3, 1876, Mrs. Crouch, a farmer's wife, was out doing some of her daily chores including the making of soap when she reported seeing chunks of meat ball from the sky. She had taken about 40 steps from her house when the meat started slapping the ground around her. Quickly joined by her husband, they believed this was a dramatic sign from God. Although these are the most famous eyewitnesses, several hundred other people through neighboring counties reported similar events the same day. Most of the meat was approximately 2 inches by 2 inches, with some reported all the way up to 5 inches by 5 inches. This strange phenomenon was covered by **Scientific America** and the **New York Times.** Dozens of people from different fields of medicine and science weighed in with what they thought the meat was and what caused the weird events. There were lots of ideas from hundreds of vultures throwing up meat, to atmospheric bacteria forming large colonies and falling to the ground. Dr. Allan McLane Hamilton stated that he identified the meat as being from lung tissue, of either horses or human infants. Hundreds of people also ate the meat, saying it tastes like everything from lamb to bear, horse, and even human. Several of the samples still exist in museums like the Smithsonian but are unable to be tested due to the age of the specimens. Most experts agree the samples look like muscle tissues and cartilage, whatever animal it could possibly be from.

Speculative Biology Analyses: This event is very similar to what we see in ocean events called whale falls, when a giant whale dies on the upper layers of the ocean, slowly falling to the bottom. In this case, it may be one of the giant manta rays or jellyfish, soft bodied and slower moving than some of the faster creatures of the upper atmosphere. When they die, as they fall back to earth, they may shred to pieces as they hit terminal velocity, with their muscles no longer locking their bodies into position at higher speeds. A creature in the ocean, often called a jelly ball, is often noted for having a mammalian taste similar horse or even bear, so a gelatinous creature from the stratosphere may also share a similar characteristic that their flesh could trick the pallets of these people into thinking that it was a mammal flavor.

Title: Eggs From Above

Location: Village of Kivalina, Alaska

Date: 2010

Witnesses: Villagers of Kivalina

Encounter: The Alaskan Village of Kivalina had a mysterious orange goo appear all over the town and shoreline. Upon further investigation under a microscope, the slimy orange goo appeared to be millions of microscopic eggs with their fatty sacks attached, but many of the experts still have not figured out what animal the eggs belong to. Janet Mitchell, City Administrator for Kivalina, says: ***"It was found several miles inland in the freshwater Wulik River. The orange material turned gooey and gave off a gaseous odor. When scooped out of the ocean, the substance had no odor and was light to the touch, with the feel of baby oil."***

The eggs are found all over the town including rooftops, rain collection buckets, sidewalks, driveways, in the local streams and the river. The locals estimate the eggs were easily in excess of thousands of gallons. NOAA scientist Jeep Rice stated that after careful microscopic examination, he cannot ID the species the eggs belong to specifically but can say with confidence they belong to some kind of invertebrate. While the city has never had the eggs return in mass numbers, they have reported some streams and creeks in the area turning orange for several days at a time.

Speculative Biology Analyses: We theorize that even though the adults of these creatures live in the upper atmosphere, they still may return to earth and the water to spawn and reproduce. Several gigantic ocean creatures, like the oceanic sunfish lay their spawn as millions - if not billions - of microscopic eggs. The creatures spawning in this event may not be tons of tiny invertebrates. Rather, it may be several large invertebrates spawning all at once.

Title: The Blob

Location: Philadelphia, Pennsylvania

Date: 1950

Witnesses: Four Local Policeman

Encounter: In the 1950s, four on duty police officers came upon the scene of a truly otherworldly situation. The four officers walked upon what would appear to be a 6-foot diameter, 1-foot thick, domed disk-shaped creature that was completely made of quivering jelly. After a while, the officers decided they should pick it up to bring it to the precinct. As they tried to pick it up, the disk dissolved into an odorless, sticky slime. This event was later blamed on the nearby Philadelphia Gas Works Company. This was also the inspiration for the movie "The Blob".

Speculative Biology Analyses: This disc could be anything. Some residual egg sacks from some insects act similarly, making it possible it is an egg sack from one of the atmospheric creatures. Another possibility is that this is a chunk of a larger creature, like a jellyfish, which are very soft and once disturbed can quickly dissipate. The way the officers described the active movement of the creature and lack of gaseous smell, this appears to be a living creature. It would potentially fit the description of a young creature that got knocked down, quickly dying after it hit the ground.

Art Credit: Starcruiser Studios

Title: Gelatinous Rain

Location: Oakville, Washington

Date: Several Dates in 1994

Witnesses: Oakville Residents

Encounter: On August 7, 1997, it started raining and jelly fell from the skies over Oakville. Small flecks, about the size of rice grains rained over the town at amazingly high volumes. It became visible over the ground, windshields, and roofs of local buildings. Again, on August 19, 1994, it rained once more. This time the jelly fell in small clear blobs, almost circular in shape. A local resident named Sunny collected some of the strange material, which would later put her in the hospital. Several other residents reported becoming sick after contact with a mysterious blob. Symptoms ranged from mild nausea and dizziness all the way to full blown immobility and severe pain. Some evidence even shows local patients had died after ingesting the strange material.

Speculative Biology Analyses: If these creatures are in fact related to jellyfish, they may have also developed the strong venom and hair trigger stinging cells of their oceanic cousins. Even after death, many species of jellyfish can still sting, even with just small chunks of their body, causing people to be hospitalized and even die. There are also cases of animals like dogs and cats dying after ingestion of jelly pieces. If these creatures of the upper atmosphere have similar stinging cells or venom and chunks of their body rained down, it would make sense that their corpses would have the ability to make people very sick, including hospitalizations and even death.

How Are Living UFO's Affecting the Abduction Phenomenon

o Missing 411

Living UFO's could hold some of the puzzle pieces for the missing 411 phenomena and Alien Abduction cases. The missing 411 phenomena was first described by famed researcher David Paulides. His documentation shows that thousands of people each year go missing in unusual circumstances. Most missing 411 cases fall within these criteria; a person goes missing in state or federal land, the person is normal and well experienced in the outdoors, bodies are usually never found, tracking dogs find little to no trails, and victims are often found far from the point of going missing. But what reasoning could exist for these upper atmosphere creatures to pray upon humans in this matter? Also, why mostly in the wilderness and not near urban areas? These creatures mostly avoid highly dense populations, likely due to light pollution and electromagnetic disturbances. These creatures most likely use bioluminescence to communicate with each other and the high levels of light pollution could heavily inhibit their ability to communicate. It would be similar to standing next to a concert speaker and trying to talk to someone next to you, which is almost impossible. Thus, they avoid these areas. As far as electromagnetic disturbances go, all electronics put these fields off, and they can be extremely disorientating to living beings. Sea turtles, birds, and sharks are some examples of animals that can become lost and even accidentally kill themselves due to these electromagnetic fields. Many animals use these fields to navigate, even if they have other organs like eyes to help them. Even with eyes, they can become easily stranded without the additional electromagnetic navigation. In relation to the missing 411 phenomena, these creatures, if they are predating on humans, would use these areas to look for food. These beings are often seen above forest and mountain ranges, with some types, like the living clouds, seen using scavenger-like behaviors. People in the missing 411 are often described to have scent trails that disappear, almost like they were picked up off the ground. As far as their belongings, some can be found some distances away. These entities have been observed to regurgitate any objects they cannot digest in piles. It is quite possible that some of the thousands of people that are reported to be part of the missing 411 phenomena are the prey of these atmospheric creatures. This would explain some of the strangeness around these poor people's disappearances and why the federal government gets involved with these people that are lost in our wilderness.

o Stealing Our Fertilizer

There are other roles that humans could play in the ecosystem of upper atmospheric life other than food.

Humans could be a roundabout way of gaining access to very desirable food sources or could play a larger role in these entities' environment. They could potentially be a source of hosts for a very complex reproductive cycle. In the sixties and seventies when alien abductions were becoming more common and gaining more of the media spotlight, the Men in Black (MIB) would come to people that had claimed to experience an abduction with a list of hundreds of questions. Among the oddest was the common question "Were there any nitrites in the vehicle with you?". The MIB always made the point to ask this question very quickly, and no matter the answer, try to skip over it as fast as possible with little explanation. This question pops up everywhere in the field with both large and small cases. Even in the Betty and Barney Hill case. In the Hill case in 1961, the couple was heading back from a spur of the moment trip to Niagara Falls. In short, they were stopped by a UFO that landed near their parked car, where they were greeted by extraterrestrials. They were both taken into the craft where they both had dramatically different events happen to them. Even though both hills described the Aliens and being very similar in looks, they had vastly different experiences. Betty was shown books and star maps and had a very enjoyable time, while Barney was tortured and molested. His experiences caused him trauma for the rest of his life. After these events, the MIB came to visit the Hills and as previously stated they proceeded to ask Barney was "Were there any nitrites in the vehicle with you?", Barney asked what a nitrite was and the MIB told him what all they could be. They could be found in processed food, chemicals, and fertilizer. He thought about the question and answered that yes, there was about four hundred pounds of fertilizer in the trunk. Upon hearing this, the MIB asked if he knew where the fertilizer was and he answered no, it had disappeared. Seeming satisfied with this answer, the MIB would not go back on this line of questioning. There were many other cases like this including a truck full of hotdogs and its driver who were involved with an abduction and every hotdog went missing. Why would aliens be interested in nitrites when they were extraterrestrials? They would have much easier ways to gain access to these resources unless they are earth born creatures wanting these nutrients for food. In the Hill abduction, they reported being followed by a light before the main event happened, and when it happened, they had dramatically different experiences. The Hill's were an interracial couple. Barney was the postmaster in his hometown in a time in American history where a black man in a high-ranking position lived a dangerous life due to racism. He lived a very stressful life due to having a high stakes job and a wife who was white in a time where black people could be beaten to death for much less. Betty, on the other hand, lived a very happy life and she and her sister loved the idea of UFO's and even meeting aliens. The way things unfolded for their abduction; their mindsets may have played a heavy role in how events unfolded. It is very similar to someone that takes hallucinogenic mushrooms for a trip. Mindset is everything when doing hallucinogens and thousands of species of mushrooms produce the chemical psilocybin, which is a psychoactive compound. Mushrooms have psilocybin in their flesh, spores, and they can even shed them in the surrounding area. These upper atmospheric creatures may be cousins to our ground-based fungus and likely have very similar abilities, like hallucinogenic dust that can cause people to have a psychedelic trip. Since the Hills described that a light was following them before the encounter, this could be these creatures getting a sense of how much food is available. When it is enough, an encounter takes place. With the Hills, the target was the four hundred pounds of fertilizer, which they seem to have taken. The encounter is caused by the creature getting very close to them, and as the Hills were having a psychedelic trip, it gathered its food and left at its convenience. The Hills slowly came out of their trip on their way home, which is another common thing with mushroom trips. But why are these animals targeting nitrites? A creature that is capable of fast flight like these would need a food source that has incredibly high nutrients and is not super heavy. So, for a fungus, a large number of nitrites would do just that. This same type of phenomena could also explain why UFOs are so interested in nuclear power plants and nuclear weapon stores. There are dozens of species of fungus that feed off nuclear waste and could be looking at these as potential food sources.

o Which is Right?

This book has presented speculative biology for UFO encounters and the theories behind their actions. There are so many theories for their behaviors and their purpose this close to the surface of the earth. Are these gentle giants of the skies above low flying scavengers of the dead, or predators that may snatch us from the dark woods? The answer is... we don't know. There is a very complex ecosystem above our heads, and whether we realize it or not, we are a part of it. Our actions are affecting a rarely researched ecosystem that may be home to earth's oldest and largest species to have ever existed. I'm personally very excited to see further research into this new field of biology. The goal of this short book is to open you to the possibility of biological entities that are flying above us with grace and a dazzling light show. Like the Blue Whales of the deepest seas, we should feel honored to share the earth with these creatures and should not fear them.

Next time you see light in the night sky, may you remember this book and look at them with curiosity and awe.

Contact Us!

Have you seen a living UFO and want to report it?

Contact through the following-

Emal:

livingufo@gmail.com

Facebook Page:

Living UFOs, The Ocean Above our heads.

Cryptids Of the Corn Podcast

Visit Our Website:

www.cryptidsofthecorn.com

Contact Us:

cryptidsofthecornpodcast@gmail.com

Patreon:

https://www.patreon.com/cryptidsofthecorn

Citation

Artangel. (n.d.). https://www.artangel.org.uk/witness/ufo-sightings-from-around-world/

Binnall, T. (2019, August 12). *"snake-like" UFO seen in New York State*. Coast to Coast AM. https://www.coasttocoastam.com/article/snake-like-ufo-seen-in-new-york-state/

Constable, T. J. (1978). *Sky creatures*. Pocket Books.

Editor, C. (2019, July 27). *Video: Snake-like ufos*. Coast to Coast AM. https://www.coasttocoastam.com/article/video-snake-like-ufos/

Home: Library of Congress. The Library of Congress. (n.d.). https://www.loc.gov/

The Lincolnite. (2023, March 8). *UFO spotter captures "Massive bright light" in Sky over Lincoln*. https://thelincolnite.co.uk/2023/03/ufo-enthusiast-spots-massive-bright-light-in-sky-over-lincoln/

Meszaros, J. (n.d.). *Giant space clams- Nevada*. Giant Space Clams- Nevada. http://statecryptids.blogspot.com/2017/04/giant-space-clams-nevada.html

NPR. (n.d.). *Breaking news, analysis, Music, Arts & Podcasts*. NPR. https://www.npr.org/

Oakville blobs: In 1994, Mysterious gelatinous goo rained down on Washington. IFLScience. (2024, April 9). https://www.iflscience.com/oakville-blobs-in-1994-mysterious-gelatinous-goo-rained-down-on-washington-73717

Pilots flash lights at UFO and it flashes back. UFOHolic. (2020, March 31). https://ufoholic.com/pilots-use-plane-lights-to-communicate-with-supersonic-ufo-and-it-flashes-back/

Researcher, byLee L. U., & Researcher, L. L. U. (n.d.-a). *TIC TAC shaped UFO sighting*. UFO Sightings Footage UK UFO Blog - Latest UFO Sightings And UAP Disclosure. https://www.ufosightingsfootage.uk/2022/06/tic-tac-shaped-ufo-sighting.html

Researcher, byLee L. U., & Researcher, L. L. U. (n.d.-b). *White UFO cloud changing shape in detail*. UFO Sightings Footage UK UFO Blog - Latest UFO Sightings And UAP Disclosure. https://www.ufosightingsfootage.uk/2021/11/white-ufo-cloud-changing-shape-in.html

Stewart, D. B. (2020, February 24). *The Giant Flying Manta-Ray of Provo Utah*. Utah Stories. https://utahstories.com/2019/04/the-giant-flying-manta-ray-of-provo-utah/

Strickler, L. (1970, January 1). Phantoms & Monsters: Pulse of the Paranormal. https://www.phantomsandmonsters.com/2021/06/

Tillman, S. (2015, July 28). *UFO / Ovni: Massive UFO sightings "Bogota colombia March 2011" (eng / ESP) - video dailymotion*. Dailymotion. https://www.dailymotion.com/video/x2zids5

Unexplained Mysteries. (2020, November 6). *Weird "sky squid" UFO filmed from airliner*. Unexplained Mysteries. https://www.unexplained-mysteries.com/news/337520/weird-sky-squid-ufo-filmed-from-airliner/

Wiki, C. to C. (n.d.-a). *Belorussian sky squid*. Cryptid Wiki. https://cryptidz.fandom.com/wiki/Belorussian_Sky_Squid

Wiki, C. to C. (n.d.-b). *Crawfordsville Monster*. Cryptid Wiki. https://cryptidz.fandom.com/wiki/Crawfordsville_Monster

Wiki, C. to C. (n.d.-c). *Dutch flying jellyfish*. Cryptid Wiki. https://cryptidz.fandom.com/wiki/Dutch_Flying_Jellyfish

Wiki, C. to C. (n.d.-d). *Flying rays*. Cryptid Wiki. https://cryptidz.fandom.com/wiki/Flying_Rays

Wiki, C. to C. (n.d.-e). *Manta-man*. Cryptid Wiki. https://cryptidz.fandom.com/wiki/Manta-Man

Wiki, C. to C. (n.d.-f). *New Delhi Celestial Serpent*. Cryptid Wiki. https://cryptidz.fandom.com/wiki/New_Delhi_Celestial_Serpent

Wiki, C. to C. (n.d.-g). *Qing Xian flying jellyfish*. Cryptid Wiki. https://cryptidz.fandom.com/wiki/Qing_Xian_Flying_Jellyfish

Wiki, C. to C.-A.-C. (n.d.). *Twister Worm*. Create. https://createacryptid.fandom.com/wiki/Twister_Worm

Wiki, C. to M. world. (n.d.). *Star Jelly*. Mysterious world Wiki. https://mysteriousworld.fandom.com/wiki/Star_jelly

Wikimedia Foundation. (2024, August 11). *Kentucky Meat Shower*. Wikipedia. https://en.wikipedia.org/wiki/Kentucky_meat_shower

YouTube. (n.d.-a). YouTube. https://www.youtube.com/watch?v=ryx9KSUMZi4

YouTube. (n.d.-b). YouTube. https://www.youtube.com/watch?v=YQGr5qVhCRw

Links for NASA Study

Full article: Stratosphere Biology (researchgate.net)

https://www.researchgate.net/publication/333968536_Stratosphere_Biology

Psalm 19:1

"The heavens declare the glory of God; the skies proclaim the work of his hands."

Psalm 23:4-6

Even though I walk through the valley of the shadow of death, I will fear no evil, for you are with me; your rod and your staff, they comfort me.
You prepare a table before me in the presence of my enemies; you anoint my head with oil; my cup overflows.
Surely goodness and mercy shall follow me all the days of my life, and I shall dwell in the house of the Lord forever.

Made in United States
Troutdale, OR
02/01/2025